let it snow!
rosemarie jarski

Dedication
To Mum – your voice is the calm of fresh fallen snow.

Author
Rosemarie Jarski is a writer and former cable TV weather girl. She lives in drizzly London with a black cat named Snowball and a Mr Frosty Ice Lolly Maker. The ring tone on her cellphone is *Let it Snow! Let it Snow! Let it Snow!*

Acknowledgements
Special thanks to Rosemary, Clare and Gareth at New Holland for coolness under pressure.

INTRODUCTION

Last winter there was a survey of supermarkets to find out which items sold best on snowy days. The most popular panic-buys were milk, bread, snack foods, toilet paper and cat litter (to sand icy paths), but way out in front, at the top of the list was... film for cameras. Seems that when it snows, what we want more than anything else is to record it. Capture it for posterity. We all have photos that bring back memories of chattering teeth and glowing noses, soaking mittens and frozen toeses.

What is it about snow that makes it so precious, so treasurable? In this celebration of winter's icy charms, you'll find the answer. Enjoy all the delights of the frosty season, from snowmen to snow angels, skating to sleigh rides and of course, snow. Proper snow, mind. None of that sissy stuff, which falls like a sprinkling of talcum powder on the street and melts away almost before it's landed. This snow sticks, swallows Scottie dogs whole and would certainly get Good King Wenceslas's royal seal of approval.

Art critic, John Ruskin once said, "Pictures of winter scenery are nearly as common as moonlights, and are usually executed by the same order of artists, that is to say, the most incapable." Such harsh criticism could not be applied to the photographic artists included here. As you gaze at their frozen landscapes from the comfort of your armchair, spare a thought for the poor photographer braving frostbite to bring us such rare beauty.

Black and white is the perfect medium to capture the frolic architecture of the snow, whether it be the sensual folds of a snowdrift or the crystalline perfection of a single snowflake. Light, texture, form and contrast are highlighted. The best photographic portraits of winter are monochrome in the same way as the best photographic portraits of people are. They focus on what is essential without the distraction of colour.

Especially evocative are pictures of snow by night. Glimmering nocturnal light infuses a snowy landscape with an otherworldliness, a sense of mystery and magic where the mind is free to wonder and reflect. In Japan, where snow is venerated as reverently as spring blossom, formal snow-viewing parties are arranged in winter, to contemplate the snow by moonlight. Watching the silvery moon dance over the snow and make it sparkle like crushed diamonds is a truly enchanting experience.

"Winter makes crystals even of ink," wrote Walter de la Mare. Complementing the photographs are words that span centuries and cultures, and reflect the myriad moods of winter, from comic to melancholic. As well as winter evergreens like Robert Frost's snowy woods and Shakespeare's icicles hanging by the wall, there are less familiar delights, like an updated version of *Jingle Bells* that will strike a chord with anyone who's ever driven through snow in an old jalopy.

Every word, every picture celebrates the sorcery of snow. What a magician Mother Nature is! With a single wave of her wand, she transforms town and countryside alike. It's the ultimate makeover. Grime is covered, jagged edges

are smoothed, and the cacophony of traffic is silenced. The world emerges purified, fresh-minted. For once, your neglected garden and clapped-out car look as good as your neighbours'. The city looks safer under a security blanket of snow; indeed, it *is* safer. Crime figures plummet on snow days. Tell-tale footprints in the snow make catching fleeing criminals a breeze.

Mother Nature works her magic on people, too. People are nicer when it snows. Like a love affair or a war, a birth or a death, a snow blizzard stops us in our tracks and forces us to take stock. Modern hi-tech society grinds to a standstill. We abandon our automobiles, take to the streets and actually notice the existence of others. We might shovel the driveway of an elderly neighbour, jump-start a stranger's broken-down car or simply smile at a passer-by. Little acts of goodwill such as these are actually what bind us together. Under snow, a city of isolated individuals becomes a real community. The public park, ordinarily the domain of self-absorbed joggers and solitary dog walkers, turns into a playground for fun-seeking families and groups of friends.

All this bonhomie makes winter the perfect time to find romance. A slip on the ice, a flat car battery, a burst water pipe, can bring a knight-in-shining-snowsuit charging to the rescue of a damsel-in-distress. The seductive quality of snow has been ruthlessly exploited by Hollywood. In *Sleepless in Seattle*, would Tom have won the fair Meg's heart if he'd wooed her not in a swirl of snowflakes but a light drizzle?

Frothy and fluffy it may be, but the effect of snow upon us is profound. Poet, Anne Sexton, writes, "I am younger each year at the first snow. When I see it, suddenly, in the air, then I am in love again and very young and I believe in everything." Snow restores the sense of wonder we had as kids. It melts away cynicism. You can't be postmodern about snow. All you can do is pull on your hat and gloves, grab a sledge and dash out to the nearest park.

"Snow is a sledding slope back to childhood," according to John Greenleaf Whittier. Snow reawakens feelings that reach right back to our earliest days, and acts as a powerful trigger to memory. Former US president, Ronald Reagan, tragically stricken with Alzheimer's disease, was given a simple Christmas gift of a snow-globe by his daughter. As he shook the tiny winter wonderland, she saw her father smile. In his blank eyes there appeared a sudden spark of recognition.

Such is the miraculous power of snow. It taps into something fundamental in our psyche. We remember our first snow like our first kiss. We vividly recall being six years old and seeing snow sashaying past our window, hearing the crunch of it underfoot, tasting its tingly coldness on our tongue.

Memories of those chilly childhood winters we all thought we had are sure to be rekindled here. So, without further ado, let's go on with the snow. Come closer, breathe softly over the frosted pane, say the magic words, "Let it Snow" and you'll be transported to the Crystal Kingdom.

FIRST SNOW

The first fall of snow is not only an event but it is a
magical event. You go to bed in one kind of a world
and wake up to find yourself in another quite different,
and if this is not enchantment, then where is it to be found?

J. B. Priestley

I can never remember whether it snowed for six days
and six nights when I was twelve or whether it snowed
for twelve days and twelve nights when I was six.

Dylan Thomas

ANATOMY OF A SNOWFLAKE

Of all the wonders of nature, the snowflake is surely one of the most beautiful and varied. Little was known about these exquisite crystals until the late 19th century when a farmer from Vermont named Wilson A. Bentley began investigating them. "Snowflake" Bentley was the first person to photograph snowflakes and he dedicated 50 years of his life to their study.

A snowflake is born when frozen water vapour falls from clouds in the form of ice crystals.

Scientists estimate 10,000,000,000,000,000,000,000,000,000,000 snowflakes have fallen to earth since it was formed. (Who counted?)

No two snowflakes are alike (how do they know?), but they do fall into seven basic types: star, needle, hexagonal, dendrite, irregular, column and capped column.

A snowflake falls at a rate of about 3 miles an hour, about six times slower than a raindrop.

THE FROLIC ARCHITECTURE OF THE SNOW

At this season I seldom had a visitor. When the snow lay deepest no wanderer ventured near my house for a week or a fortnight at a time, but there I lived as snug as a meadow mouse…

   As I went forth early on a still and frosty morning, the trees looked like airy creatures of darkness caught napping; on this side huddled together, with their gray hairs streaming, in a secluded valley which the sun had not penetrated; on that, hurrying off in Indian file along some watercourse, while the shrubs and grasses, like elves and fairies of the night, sought to hide their diminished heads in the snow.

   I frequently tramped eight or ten miles through the deepest snow to keep an appointment with a beech tree, or a yellow birch, or an old acquaintance among the pines.

Henry David Thoreau

I made a snowman and my brother knocked it down
and I knocked my brother down and then we had tea.

Dylan Thomas

THE WHITE STUFF

Cool Facts About Snow

Snow is not white. The whiteness we perceive comes from light reflecting and refracting from the surface of the crystals.

The heaviest fall of snow in a single day was at Silver Lake, Colorado, on 14–15th April 1921. In total, 74 inches of snow fell.

A bitterly cold winter in 1632 enabled the Swedish army to invade Germany by marching across the frozen Baltic Sea.

If you ever find yourself buried under an avalanche, survival experts advise that you first clear the snow surrounding you then spit into it. Whichever way the saliva goes, up will be in the opposite direction.

The Ice Hotel in Jukkastjärvi, Lapland, is made of 3,000 tons of ice and 130,000 cubic metres of snow. Features include an ice bar, an ice-screen cinema and an ice chapel where you can get married. (You may be relieved to hear that the toilets are one of the few items not made of ice!)

A rare snowfall in Renaissance Italy in January 1484 so excited the Florentine ruler, Piero de'Medici, that he commissioned a snowman to grace his garden. The sculptor he hired was one Michelangelo.

Chocolate snow? When a dust storm coincided with a snowstorm over Mount Hotham in Victoria, Australia, in 1935, the result was chocolate-coloured snow!

The Snow Crystals Museum in Asahikawa, Hokkaido, Japan, is devoted exclusively to all things snowy and is a must for snow-addicts.

JINGLE BELLS

Dashing through the snow,
In a one-horse open sleigh;
O'er the fields we go,
Laughing all the way;
Bells on bob-tail ring,
Making spirits bright;
What fun it is to ride and sing
A sleighing song tonight.

Jingle bells, jingle bells,
jingle all the way;
O what joy it is to ride
In a one-horse open sleigh.
Jingle bells, jingle bells,
jingle all the way.
O what joy it is to ride
In a one-horse open sleigh.

James Pierpont

RUSTY CHEVROLET

Dashing through the snow
in my rusty Chevrolet.
Down the road I go,
sliding all the way;
I need new piston rings,
I need some new snow tires,
My car is held together
by a piece of chicken wire!

Oh, rust and smoke,
the heater's broke,
the door just blew away.
I light a match to see the dash
and then I start to pray-ay.
The frame is bent,
the muffler went,
the radio's okay.
Oh, what fun it is to drive
this rusty Chevrolet!

DeCaire & Potila

To
boggan?
or not
to boggan?
That is the question.

Roger McGough

TOBOGGAN

Down from the hills and over the snow
Swift as a meteor's flash we go,
Toboggan! Toboggan! Toboggan!
Down from the hills with our senses lost,
Jealous of cheeks that are kissed by the frost,
Toboggan! Toboggan! Toboggan!

Down from the hills, what an awful speed!
As if on the back of a frightened steed,
Toboggan! Toboggan! Toboggan!
Down from the hills at the rise of the moon,
Merrily singing the toboggan tune,
"Toboggan! Toboggan! Toboggan!"

Down from the hills like an arrow we fly,
Or a comet that whizzes along through the sky;
Down from the hills! Oh, isn't it grand!
Clasping your best winter girl by the hand,
Toboggan! Toboggan! Toboggan!

Benjamin Franklin King

MORE SNOW FALLING

One white hush the whole day.
No wind. Just endless in-
exorable cliché,
the same old stuff again

and again; We need old stuff
sometimes. As in liturgy.
Or a declaration of love.
No frills, just infinity.

Peter Kane Dufault

SNOW JOKES

✼ How does Good King Wenceslas like his pizza?
Deep pan, crisp and even.

✼ Where do snowmen go to dance?
Snowballs.

✼ Two Eskimos sitting in a kayak were chilly,
but when they lit a fire in the craft it sank—
proving once and for all that
you can't have your kayak and heat it, too.

✼ What do snowmen eat for breakfast?
Snowflakes.

* What do you get if you cross a snowman and a shark?
Frostbite.

* How do you get milk from a polar bear?
Rob its fridge and run like mad.

* Knock knock.
  Who's there?
  Wenceslas.
  Wenceslas who?
  Wenceslas train home?

* What did Jack Frost say to Frosty the Snowman?
Have an ice day.

NO BUSINESS LIKE SNOW BUSINESS

Hollywood Snow

Southern California is not renowned for its snow. When movie mogul, Louis B. Mayer, wanted a white Christmas he ordered tons of "movie snow" from the studio to be spread over the lawns of his beach house in Santa Monica. "Movie snow" can take various forms depending on the type of snow required:

✴ rock salt – fallen snow on the ground

✴ shredded chicken feathers blown by fans – light snow in the air

✴ gypsum – heavy snow

✴ white-coated cornflakes – crunchy, fresh fallen snow

✴ table salt – close-ups on actors' clothing and hair

The classic movie, *It's A Wonderful Life*, was pioneering in its depiction of snow. Director, Frank Capra, invented a new technique which mixed foamite (the chemical used in fire extinguishers) with soap and water. The resulting snow was very realistic. "Tastes swell, too," said Jimmy Stewart, "but unless you work fast, you're apt to start frothing at the mouth."

WINTER BLOSSOM

Sei Shōnagon was a lady-in-waiting at the Heian court in 10th-century Japan. As was the practice, she kept a "pillow book" in which she jotted down her thoughts and observations on everything under the sun—and snow…

Things That Fall from the Sky
Snow. Hail. I do not like sleet, but when it is mixed with pure white snow it is very pretty. Snow looks wonderful when it has fallen on a roof of cypress bark. When snow begins to melt a little, or when only a small amount has fallen, it enters into all the cracks between the bricks, so that the roof is black in some places, pure white in others—most attractive.

Unsuitable Things
Snow on the houses of common people. This is especially regrettable when the moonlight shines down on it.

To Meet One's Lover
In the winter, when it is very cold and one lies buried under the bedclothes listening to one's lover's endearments, it is delightful to hear the booming of a temple gong, which seems to come from the bottom of a deep well.

WHEN ICICLES HANG BY THE WALL

When icicles hang by the wall,
And Dick the shepherd blows his nail,
And Tom bears logs into the hall,
And milk comes frozen home in pail,
When blood is nipp'd, and ways be foul,
Then nightly sings the staring owl
Tu-whit, *to-who*—a merry note,
While greasy Joan doth keel the pot.

When all aloud the wind doth blow,
And coughing drowns the parson's saw,
And birds sit brooding in the snow,
And Marion's nose looks red and raw,
When roasted crabs hiss in the bowl,
Then nightly sings the staring owl,
Tu-whit, *to-who*—a merry note,
While greasy Joan doth keel the pot.

William Shakespeare

IN THE BLEAK MIDWINTER

In the bleak midwinter
Frosty wind made moan,
Earth stood hard as iron,
Water like a stone;
Snow had fallen, snow on snow,
Snow on snow,
In the bleak midwinter,
Long ago.

Christina Rossetti

REQUIESCAT

Tread lightly, she is near
Under the snow,
Speak gently, she can hear
The daisies grow.

Oscar Wilde

WINTRY WISDOM

Wherever snow falls there is usually civil freedom.
Ralph Waldo Emerson

The snow doesn't give a soft white damn whom it touches.
e.e. cummings

Growing up in a place that has Winter you learn to avoid self-pity. Winter is not a personal experience – everybody else is as cold as you – so you shouldn't complain about it too much.
Garrison Keillor

No snowflake in an avalanche ever feels responsible.
Stanislaw J. Lec

The wisdom of a single snowflake outweighs
the wisdom of a million meteorologists.
Sir Francis Bacon

> Ice-skating over thin ice our safety is in our speed.
> Ralph Waldo Emerson

A man shovels snow for the same reason he climbs a mountain –
because it's there.
Nathan Nielsen

The dirty puddle used to be pure snow. I walk by it with respect.
Stanislaw J. Lec

WINTRY WISECRACKS

I used to be Snow White, but I drifted.
Mae West

I'm as pure as the driven slush.
Tallulah Bankhead

Cross-country skiing is great if you live in a small country.
Steven Wright

Generally speaking, the poorer person winters where he summers.
Fran Lebowitz

The coldest winter I ever spent was a summer in San Francisco.
Mark Twain

STOPPING BY WOODS ON A SNOWY EVENING

Whose woods these are I think I know.
His house is in the village though;
He will not see me stopping here
To watch his woods fill up with snow.

My little horse must think it queer
To stop without a farmhouse near
Between the woods and frozen lake
The darkest evening of the year.

He gives his harness bells a shake
To ask if there is some mistake.
The only other sound's the sweep
Of easy wind and downy flake.

The woods are lovely, dark and deep,
But I have promises to keep,
And miles to go before I sleep.
And miles to go before I sleep.

Robert Frost

WINTER DREAMS

Snow is a symbol of purity, beauty, cleansing and innocence.
If you dream of snow this is what it means…

✳ snow – cleansing, purity

✳ snow falling – interesting changes will occur in your life

✳ snowdrift – expect good fortune in money or love

✳ snow melting – crystallization of an idea; softening of the heart

✳ iceberg – beware what lies beneath the surface

✳ icicles – fears and anxieties will soon melt away

✳ dripping icicles – do not part with your money

Remember: sometimes an icicle is just an icicle.

A WINTRY LETTER

24 December 1874

Dear Mrs Sitwell

Outside, it snows thick and steadily. The gardens before the house are now a wonderful fairy forest. And O, the whiteness of things, how I love it, how it sends the blood about my body! Maurice de Guérin hated snow; what a fool he must have been! Somebody tried to put me out of conceit with it by saying that people were lost in it. As if people don't get lost in love, too, and die of devotion to art; as if everything of worth were not an occasion to some people's end.

What a wintry letter this is! Only I think it is winter seen from the inside of a warm greatcoat. And there is, at least, a warm heart about it somewhere. Do you know, what they say in Xmas stories is true. I think one loves their friends more dearly at this season.

Ever your faithful friend,

Robert Louis Stevenson

BRRRRR..... HOW COLD IS IT?

It's so cold that if I sat on a tack, I wouldn't notice it until the spring thaw.
W.C. Fields

My room's so cold that every time I open the door the light goes on.
Steve Martin

It's so cold in New York that the Statue of Liberty is holding the torch *under* her dress.
David Letterman

It was so cold I saw a politician with his hands in his own pockets.
Bob Hope

I did a picture in England one winter and it was so cold I almost got married.
Shelley Winters

It's so cold that I have to use an ice pick to get comfortable on my waterbed.
Groucho Marx

It's so cold that cows are giving cream by the scoop.
George Burns

The frost was severer than ever in the night as it even froze the Chamber Pots under the Beds.
The Reverend J. Woodforde, 1790

THE HILLS IN SNOW

It is marvellous weather but immensely cold – everything frozen solid –
milk, mustard, everything. Yesterday I went out for a real walk.
I climbed to the top of the hills. Wonderful it is to see the footmarks
on the snow – beautiful ropes of rabbit prints, trailing away over the brows;
heavy hare marks; a fox so sharp and dainty, going over the wall:
birds with two feet that hop; very splendid straight advance of a pheasant;
wood-pigeons that are clumsy and move in flocks; splendid little leaping
marks of weasels coming along like a necklace chain of berries;
odd little filigree of the field-mice; the trail of a mole – it is astonishing what a
world of wild creatures one feels about one, on the hills in snow.

D.H. Lawrence

THE MORE IT SNOWS

The more it snows
  (Tiddely pom),
The more it goes
  (Tiddely pom),
The more it goes
  (Tiddely pom),
  On snowing.
And nobody knows
  (Tiddely pom),
How cold my toes
  (Tiddely pom),
How cold my toes
  (Tiddely pom),
  Are growing.

A.A. Milne

THE FALLOW DEER AT THE LONELY HOUSE

One without looks in tonight
Through the curtain-chink
From the sheet of glistening white;
One without looks in tonight
  As we sit and think
  By the fender-brink.

We do not discern those eyes
Watching in the snow;
Lit by lamps of rosy dyes
We do not discern those eyes
  Wondering, aglow,
  Fourfooted, tip-toe.

Thomas Hardy

SNOW HAS FALLEN

The year is nearly over. Snow has fallen, and everything is white. It is very cold. The forlorn wind scarcely breathes. I love to close my eyes a moment and think of the land outside, white under the mingled snow and moonlight – white trees, white fields – the heaps of stones by the roadside white – snow in the furrows. *Mon dieu*! How quiet and how patient! If he were to come I could not even hear his footsteps.

Katherine Mansfield

A SNOW GLOBE OF PROVERBS

You can do nothing about governments and winter.
Slovenian

Better the cold blast of winter than the hot breath of a pursuing elephant.
Chinese

Before you love, learn to run through snow without leaving footprints.
Turkish

No snowflake ever falls in the wrong place.
Japanese

One kind word can warm three winter months.
Japanese

> A fat woman is a blanket for the winter.
> Arabian

Out of snow, you can't make cheesecake.
Jewish

> The church is near, but the way is icy.
> The tavern is far, but I will walk carefully.
> Ukranian

THE BEAUTIFUL SNOW

O the snow, the beautiful snow,
Filling the sky and the earth below!
Over the house-tops, over the street,
Over the heads of the people you meet,
  Dancing,
    Flirting,
      Skimming along.
Beautiful snow! It can do nothing wrong.
Flying to kiss a fair lady's cheek;
Clinging to lips in a frolicsome freak.
Beautiful snow from the heavens above,
Pure as an angel and fickle as love!

James. W. Watson

UP IN THE MORNING EARLY

Cauld blaws the wind frae east to west,
The drift is driving sairly,
Sae loud and shrill's I hear the blast—
I'm sure it's winter fairly.

Up in the morning's no for me,
Up in the morning early!
When a' the hills are cover'd wi' snaw,
I'm sure it's winter fairly!

The birds sit chittering in the thorn,
A' day they fare but sparely,
And lang's the night frae e'en to morn—
I'm sure it's winter fairly.

Robert Burns

VELVET SHOES

Let us walk in the white snow
In a soundless space;
With footsteps quiet and slow,
At a tranquil pace,
Under veils of white lace.

I shall go shod in silk,
And you in wool,
White as white cow's milk,
More beautiful
Than the breast of a gull.

We shall walk through the still town
In a windless peace;
We shall step upon white down,
Upon silver fleece,
Upon softer than these.

We shall walk in velvet shoes:
Wherever we go
Silence will fall like dews
On white silence below.
We shall walk in the snow.

Elinor Wylie

SYMPHONY IN SNOW

It was one of those brilliant, glittery snows that ought to emit some glorious sound with each crystal falling to earth, something transcendent like a Bach cantata.
Lynne Sharon Schwartz

The snow sounded as if someone was kissing the window all over outside.
Lewis Carroll

It can become so cold in Siberia that your breath freezes into ice crystals and tinkles to earth with an ethereal sound they call 'the whispering of stars'.
Phyl Amison

The most delightful advantage of being bald – one can *hear* snowflakes.
R.G. Daniels

SNOW TO SPEAK

Arabs have a hundred words for camel, Italians have a hundred words for pasta, and New Yorkers have a thousand words for jerk, but how many words for snow do Eskimos have? Hotly debated for years, this mystery is finally solved by Inuit-language expert, Phil James. Try them out on the next Eskimo you rub noses with…

tlapa	powder snow	*quinyaya*	snow mixed with dung of lead dog
tlacringit	crusted snow	*slimtla*	snow crusted on top but soft below
kayi	drifting snow	*kriplyana*	snow that looks blue in the morning
tlapat	still snow	*jatla*	snow between your fingers or toes
klin	remembered snow	*dinliltla*	balls of snow clinging to Husky fur
naklin	forgotten snow	*sulitlana*	green snow
tlamo	snow falling as large wet flakes	*mentlana*	pink snow
tlatim	snow falling as small flakes	*tidtla*	snow used for cleaning
tlaslo	snow that falls slowly	*ertla*	snow teenagers use for erotic rituals
tlapinti	snow that falls quickly	*kriyantli*	snow bricks
kripya	snow melted and refrozen	*hahatla*	packages of snow given as gag gifts
tliyel	snow marked by wolves	*semtla*	partially melted snow
tliyelin	snow marked by Eskimos	*ontla*	snow on objects
blotla	blowing snow	*intla*	snow that has drifted indoors
pactla	snow packed down	*shlim*	slush
hiryla	snow in beards	*warintla*	snow used to make daiquiris
wa-ter	melted snow	*mextla*	snow used to make Margaritas
tlayinq	snow mixed with mud	*penstla*	the idea of snow
quinaya	snow mixed with Husky dung	*mortla*	snow mounded on dead bodies

ylaipi	tomorrow's snow
nylaipin	the snows of yesteryear
pritla	our children's snow
nootlin	snow that doesn't stick
trinkyi	first snow of the year
tronkyin	last snow of the year
shiya	snow at dawn
katiyana	night snow
tlinro	snow vapour
nyik	snow with flakes of widely varying size
ragnitla	two snowfalls at once, creating moiré patterns
akitla	snow falling on water
privtla	snow melting in spring rain
chahatlin	snow that sizzles as it falls on water
hootlin	snow that hisses as individual flakes brush
geltla	snow dollars
briktla	good building snow
striktla	snow that's no good for building
erolinyat	snow drifts showing imprints of crazy lovers
chachat	swirling snow that drives you nuts
krotla	snow that blinds you
tlarin	snow that can be sculpted into the delicate corsages Eskimo girls pin to their whale parkas at prom time
motla	snow in the mouth
sotla	snow in the south
maxtla	snow that hides the whole village
tlayopi	snow drifts you fall into and die
truyi	avalanche of snow
tlapripta	snow that burns your scalp and eyelids
carpitla	snow glazed with ice
tla	ordinary snow
allatla	baked snow
puntla	a mouthful of snow because you fibbed
fritla	fried snow
gristla	deep fried snow
MacTla	snow burgers
rotlana	quickly accumulating snow
skriniya	snow that never reaches the ground
bluwid	snow shaken from objects in the wind
tlanid	snow shaken and mixed with sky-falling snow
ever-tla	a spirit made from mashed fermented snow
talini	snow angels
priyakli	snow that looks like it's falling upward
chiup	snow that makes halos
blontla	snow that's shaken off in the mudroom
tlalman	snow sold to German tourists
tlalam	snow sold to American tourists
tlanip	snow sold to Japanese tourists
protla	snow packed around caribou meat
attla	falling snow that creates pictures in the air
sotla	snow sparkling with sunlight
tlun	snow sparkling with moonlight
astrila	snow sparkling with starlight
clim	snow sparkling with flashlight or headlight
tlapi	summer snow
krikaya	snow mixed with breath
ashtla	expected snow that's wagered on (size of flakes etc.)
huantla	special snow rolled into "snow reefers"
tla-na-na	snow mixed with the sound of old rock & roll from a portable radio
depptla	a small snowball, preserved in Lucite, that has been handled by Johnny Depp

SKIING EXERCISES

Visit your local butcher and pay £30 to sit in the walk-in freezer for half an hour. Afterwards, burn two £50 pound notes to warm up.

Buy a brand new pair of gloves and immediately throw one away.

Dress up in as many clothes as you can and then proceed to take them off because you have to go to the bathroom.

Secure one of your ankles to a bedpost and ask a friend to run into you at high speed.

Fill a blender with ice, hit the pulse button and let the spray blast your face. Leave the ice on your face until it melts. Let it drip into your clothes.

Drive slowly for five hours – anywhere – as long as it's in a snowstorm and you're following a bus.

Find the nearest ice rink and walk across the ice 20 times in your ski boots carrying two pairs of skis, an accessory bag and poles. Sporadically drop things.

ICE SKATING

Ice skating is a feeling of ice miles running under your blades, the wind splitting open to let you through, the earth whirling around you at the touch of your toe, and speed lifting you off the ice far from all that can hold you down.

Sonja Henie

SNOW BATHING

When I was five, my white adoptive parents found me
rolling about in the snow outside my home in Kent.
I'd gone missing minutes earlier and they'd dashed out
of the house to scour the village. When they found me,
I was lying naked in the icy whiteness with what I
thought was a rational explanation for my behaviour –
white people lay in the sun to go brown, therefore,
I reasoned, I would turn white if I 'snow-bathed'.

Sarah Obanye

ALL WHITE NOW

Snow is as much a part of the Yuletide festivities as pine needles on the carpet, rows with relatives and Alka-Seltzer™. Forget religion and presents, Christmas isn't Christmas without any snow.

Charles Dickens invented the iconic white Christmas. His formative childhood Christmases were all snowbound and he drew on these when creating *A Christmas Carol*. Published in December 1843, the book became a bestseller and so was born the myth that Christmas can be any colour you like so long as it's white.

A century later, the idea snowballed when Bing Crosby crooned a little song called *White Christmas* in the movie musical, *Holiday Inn*. Irving Berlin's simple words struck a chord in the dark days after Pearl Harbour, particularly with U.S troops posted in the sunny Pacific and longing to spend Christmas back home with loved ones.

The enduring appeal of the image is reflected in the fact that *A Christmas Carol* is Dickens's most popular work and *White Christmas* has become the world's top-selling and most frequently recorded song. Four hundred million copies have been sold, 31 million for Bing alone.

WHITE CHRISTMAS

I'm dreaming of a white Christmas
Just like the ones I used to know.
Where the treetops glisten,
And children listen
To hear sleigh bells in the snow.
I'm dreaming of a white Christmas
With every Christmas card I write.
May your days be merry and bright.
And may all your Christmases be white.

Irving Berlin

Credits

Pictures
Getty Images: pages 2; 4; 7; 8; 11; 12; 14–15; 16; 19; 20; 25; 27; 28; 31; 32; 33; 37; 38; 41; 42; 44; 49; 50; 53; 61; 62; 65; 66; 69; 72; 75; 76; 79; 80; 81; 85; 86; 90; 93; 96.
© Image 100/Royalty-Free/CORBIS: page 89.
Knebworth House and Gardens, Knebworth, Hertfordshire by Henry Lytton Cobbold
© Henry Lytton Cobbold, courtesy of www.knebworthhouse.com: page 54.
Elliott Erwin/Magnum Photos: page 57.
Craig Connor, North News & Pictures: page 45.
Topham Picturepoint: page 21.
V&A Pictures/The Victoria and Albert Museum: page 68.

Text
The author and publisher are grateful to the following for permission to reproduce quotations:

Pages 13 and 21 – Excerpts from *A Child's Christmas in Wales* by Dylan Thomas, copyright © 1954, J.M. Dent. Reprinted by permission of David Higham Associates.

Page 28 – *Poem ii* of "Poems of Snow and Ice" from PIE IN THE SKY by Roger McGough (Puffin Books) 1985. Reprinted by permission of PFD on behalf of: Roger McGough.

Page 39 – Excerpt from *The Pillow Book of Sei Shōnagon* translated by Ivan Morris (1967), reprinted by permission of Oxford University Press.

Page 51 – *Stopping by Woods on a Snowy Evening* by Robert Frost from THE POETRY OF ROBERT FROST, edited by Edward Connery Latham and published by Jonathan Cape. Used by permission of the Estate of Robert Frost and The Random House Group Limited.

Page 63 – *The More It Snows* from WINNIE-THE-POOH by A.A. Milne, copyright under the Berne Convention, reproduced by permission of Curtis Brown Ltd., London.

Page 92 – *White Christmas*, words and music by Irving Berlin © 1940, 1942 (renewed) Irving Berlin Music Corp, USA Warner/Chappell Music Ltd, London W6 8BS. Reproduced by permission of International Music Publications Ltd, All Rights Reserved.

Every effort has been made to contact the copyright holders of the material quoted; but if any has been inadvertently overlooked the necessary correction will be made in any future edition of this book.

First published in 2004 by
New Holland Publishers (UK) Ltd
London • Cape Town • Sydney • Auckland
www.newhollandpublishers.com

Garfield House, 86–88 Edgware Road
London W2 2EA
United Kingdom

80 McKenzie Street
Cape Town 8001
South Africa

14 Aquatic Drive
Frenchs Forest, NSW 2086
Australia

218 Lake Road
Northcote, Auckland
New Zealand

1 3 5 7 9 10 8 6 4 2

Text copyright, except where otherwise stated, © Rosemarie Jarski
Copyright © 2004 New Holland Publishers (UK) Ltd

Rosemarie Jarski has asserted her moral right to be identified as the author of this work.

All rights reserved. No part of this publication may be reproduced, stored in a retrieval system,
or transmitted in any form or by any means, electronic, mechanical, or otherwise,
without the prior written permission of the copyright holders and publishers.

ISBN 1 84330 743 X

Editors: Clare Hubbard and Gareth Jones
Editorial Direction: Rosemary Wilkinson
Design: Paul Wright
Production: Hazel Kirkman

Reproduction by Modern Age Repro House Ltd, Hong Kong
Printed and bound by Tien Wah Press Ltd, Singapore